www.DaveGerber.info

On Fire Leadership®

Preparing for Robots: Motivational Quotes for AI, IT and Technology Professionals

Published by Timeless Publishing
www.Timelesspublishing.com

E-mail: info@timelesspublishing.com

Printed in the United States of America

www.PreparingforRobots.com

Dedication

To all those who are pointing out the challenges that come along with overwhelming technology and want to preserve the human element for as long as possible.

To those who are using their talents with technology for the good of humankind.

To my daughter, Jessie.

Welcome Fellow Humans!

I'm your host, Dave Gerber! And you have tuned in to our new segment, Preparing for Robots! Well, in this case, this book is just one aspect of the Preparing for Robots concept that I have introduced.

Maybe you get it…this massive phenomenon will impact us at every level…personal, interpersonal, social, economic, political, environmental and so much more. We have to get in the game…and pay attention. While doing so, we are going to need a little motivation and humor…this book can hopefully provide a little of that for you.

This "Preparing for Robots!" concept is about the impact of everything digital on humans, and not just robots like from the television shows! The buzz of new tech vocabulary rings in our ears and bounces like a pinball machine in our heads…block chain, Crypto, AI, AR, VR, MR, real robots, wearable robots, 3D printing (there making bones, this is crazy), the internet of things and so much more.

This is critical to understand, robots don't fight, steal time, need meetings, have conflict, miscommunicate, act out, create unions, require breaks, have worker safety laws, they are not impulsive or irrational, require health care, sue for discrimination, have dips in productivity levels or require insurance and more…oh…and they don't feel (yet.) We do.

In many ways, the deck is stacked against us. And yet…

What is coming in our digital revolution is so huge, like a tsunami sized baseball bat to the forehead. Remember tee ball?…just sitting there waiting to be smacked. 50% of the jobs we hold will be gone in the next 10-12 years. That is 2030, right around the corner. The impact of digital technology on humans.

When I speak to groups about conflict and/or preparing for the new digital age I stress a few things.

If digital technology is going to take jobs, what are the skills that people will need to have to prove value? Innovation, conflict management, creative problem solving, emotional intelligence, discipline and higher order thinking are the main sources of ROI humans will bring with them to the table. We need to get better solutions to conflict and use these other skill sets as well, now. There are low hanging fruit solutions and we had better get to it as we

Prepare for Robots with a little more sense of urgency!

There are so many topics to discuss and questions to be answered, probably the same as the number of stars in the sky.

Maximizing our own potential and harnessing others to make it happen for them needs to be a part of our legacy, our brand, our business, our life…with ourselves and others personally, and socially. In doing so we can strive to become the best version of ourselves…that's the best we can do on our pursuit of happiness and to make ourselves economically competitive. It's my hope that this book can help be a part of that journey for you.

You can follow on my work as "The conflict healer" – My approach is unique to the planet, bringing together expertise in leadership and conflict as a long time consultant, coach, educator combined with being a black belt self-defense instructor and a certified Reiki master. I Love Ya and…Now to the quotes!

Any sufficiently advanced
technology is
indistinguishable from
magic.
~Arthur C. Clarke

1
The use of technology needs to improve human's ability to learn and be more, well, human.

2
Sometimes what you take away may be more valuable than what you add.

3

Gratitude is the root of all good things including each time the computer turns on.

4

Don't let technology harden you are steal your compassion.

5

Demonstrate the characteristics that are solely human, like compassion and joy, respect and recognition, humility and happiness. Robots will never have these qualities.

6

If you want to make a significant financial future for yourself, get involved in AI, AR, VR and the host of other fields slamming our market in the next three years.

7

The internet has enabled you to learn whatever you want for wherever you want to go.

8

Communication between humans is far more difficult than dealing with technology. Make the effort, pay attention to details and come from a place of real care.

9

Make your story come to life with the newest technology available as it will be your visual legacy.

10

You must never live with regrets.

11

The difference between the right word and the almost right word is the difference between lightning and the lightning bug.

~Mark Twain

12
Computers only give answers. Real living resides in the quality of the questions.

13
How will saying "No" to one thing in your life free up so many "Yeses."

14
It has become appallingly obvious that our technology has exceeded our humanity. On a personal level, we can still do something about it.
~Albert Einstein

15

There has never been a better time in history to be yourself, even if you only share it with a true few.

16

It is the creativity, curiosity and inertia of the human mind that will be responsible for creating something far more advanced than humans themselves.

17

The computer is going to do what you tell it. If the answer doesn't make sense, don't blame the computer.

18

If you want to put your tech based, financial future in the spotlight you need to get into AI, VR, AR or the next generation of advancements.

19

Part of the inhumanity of the computer is that, once it is competently programmed and working smoothly, it is completely honest.
~Isaac Asimov

20

If you can move people to action you are a leader.

21

Humans have the power to heal, to capture energy and use it to improve self or other's physical, emotional or spiritual lives. Technology will never be able to do this.

22

The business model hasn't changed now that technology has progressed so much. It is still about finding solutions to stakeholder needs.

23

Be sure to setup your meetings to consider the number and how your extroverts and introverts will interact with one another.

24

The only way to win is not to play.

25

There is no amount of protections in the world that will guard us from the digital criminal, but we must try everything within our power.

26

If you are in business and not talking about Information Technology you are likely missing out on so many potential clients.

27

To be successful requires one to be ethical.

28

There is so much good reading, don't get completely lost in the world of video.

29

Once technology works it's all downhill from there.

30

Motivate your colleagues to be better by consistently demonstrating high level communication skills.

31

I speak to everyone in the same way, whether he is the garbage man or the president of the university.

~Albert Einstein

32

Success is not final, failure is not fatal: it is the courage to continue that counts.

~Winston Churchill

33

If people are throwing bricks or obstacles your way the only thing to do is use them to reinforce your firewalls.

34

If you knew tomorrow would be his or her last, would you treat them any differently?

35

We find happiness when we combine our passion with the content knowledge we love.

36

Remember the distinction between change and transition. Transition is probably more difficult to deal with as it is the experience drawn out over time vs. something that changes like a light switch.

37

Replace the word risk with the word opportunity.

38

Innovation, creativity, thinking outside the box...there has never been a better time for it than now.

39

The power of the 'delete' button on your keyboard is linked to our physiology, the way we think and feel. Hit delete more often and free yourself.

40

Any problem offers you the ability to be the one who offers the solution.

41

Next time you get a new computer by a foam bat because you are going to want to beat it for doing something you probably told it to do. You will still feel better.

42

In order to truly understand ourselves we must investigate the role of our family, peers, media, school and religion on my childhood.

43

Control only what you can and let the rest go.

44

Every positive value has its price in negative terms... the genius of Einstein leads to Hiroshima.
~Pablo Picasso

45

Don't focus on yesterday so much that you ruin today.

46

It is an immediate red flag if no one is asking questions.

47

Moral outrage does not weigh more than facts.

48

It is all about ownership. From the emotional, the physical and the financial.

49

Never let emotion supersede logic.

50

I think it's very important to have a feedback loop, where you're constantly thinking about what you've done and how you could be doing it better.

~Elon Musk

51

There is a great difference between empowerment and being powerful. To misunderstand this could be deadly.

52

With ideas it is like with dizzy heights you climb: At first they cause you discomfort and you are anxious to get down, distrustful of your own powers; but soon the remoteness of the turmoil of life and the inspiring influence of the altitude calm your blood; your step gets firm and sure and you begin to look - for dizzier heights.
~Nikola Tesla

53

All opinions are not equal. It is what it is.

54

Early birds make the originals.

55

Inspire your team to simplify whenever possible.

56

When they don't blame the tech department they still might blame you anyway. That's their stuff, don't own it and it won't bother you.

57

Be careful for the meetings after the meetings.

58

Get in the habit of encouraging and teaching.

59

Help inspire someone else to become the best version of themselves.

60

Be sure to catch people doing things right and point it out.

61

Don't let idealism get in the way of realism.

62

Don't try to outsmart a computer. Focus all energies going forward on things technology does not do well.

63

If you know how to build it you will better know how to teach it.

64

Be careful of what you tell a computer to do even if you didn't mean to.

65

The toughest people never talk about being tough.

66

Freedom is everything. Don't look up one day and realize it from a place of not having it.

67

While it may seem distant in society today, education is designed to have you learn through questioning and challenge the narrative.

68

Never underestimate the power of a simple good deed.

69

Don't let your cell phone kill your reflective moments.

70

If humans don't always think straight, what do you really expect computers to do?

71

You can now bring an amazing idea to market in days.

72

Knowledge is power, the real question is what will you do with it?

73

Learn the names of the gatekeepers in your organization, including the people who clean the building.

74

Tell them, tell them what you just told them and then tell them again if you really want them to remember.

75

Look for ways to disconnect from tech and connect with nature.

76

I never see what has been done; I only see what remains to be done. ~Buddha

77

When in doubt, read and learn more.

78

Life is about action. Sometimes we find action leads to the wrong answers which then helps us to get to the right place faster. It starts with motion.

79

You must be able to do more together than on your own or get a new partner.

80

Practice when it's easy so you can come out on top when things are difficult.

81

Someone who loses but retains their character and moral compass wins in most everyone's perspective.

82

The first country to discover the highest level of AI will be dominant in almost everything. How can you be a part of this race?

83

Search for real intimacy and connection with humans.

84

Set a destination but be intuitive enough to know when to find another pathway.

85

The only right way is the way that works.

86

Be sure that whether you are waking up or going to bed, your phone isn't the last thing that impacted your head.

87

In any project it's all about your ability to impact the Critical Path.

88

The key to immortality is first living a life worth remembering.
~Bruce Lee

89

Start by doing what is necessary; then do what's possible; and suddenly you are doing the impossible.

~Francis of Assisi

90
Be a 'Hope Dealer.'

91

The Kaizen way says start small, just start. If you want to clean the desk, start with moving on staple, one rubber band, one out of place book. This thinking can help you get started with anything. Just start.

92
You can discover more about a person in an hour of play than in a year of conversation.
~Plato

93

It seems that no matter how fast technology goes it is still not fast enough.

94

Computers and technology now is truly about the integration with our life, our environment and our bodies.

95

Become a beginner at something. You don't have to tell anyone. Whether you do or do not share, experiencing humility and growth at the same time is priceless.

97

Practice only makes perfect if you are doing it right.

98

Each week think about someone new you could help and make something happen with your newfound care.

99

"Spiritual Stoicism" is the addition of personal growth, understanding self and leveling up beyond just turning your emotions off.

100

In order to carry a positive action we must develop here a positive vision.

~Dalai Lama

101
Gratitude is the attitude we don't want to keep to ourselves.

102
The more you trust your intuition the more it will speak to you.

103
We have a society of over-sharing, be careful what you post. Live the difference between public and private life.

104
The difference between a successful person and others is not a lack of strength, not a lack of knowledge, but rather a lack of will.
~Vince Lombardi

105
When you decode something someone says, be sure you didn't interpret it the worst way possible.

106
If you didn't create the policy, can't influence or change the policy, what is making you continue a conflict over the policy?

107
Be sure to add a little art to your life, the kind created by humans.

108
Find your trustworthy advocates that believe in you and can keep their mouth shut.

109
When learning something new use the method of teaching it back to someone as your retention rate will be near 90%.

110

Accept the challenges so that you can feel the exhilaration of victory.

~George S. Patton

111

If you are not taking any risks you will never see the big rewards you really want.

112

Be prepared, work to the best of your ability and you will have less worry.

113

High Performance Teams require shared risk taking, high levels of technical competence and diverse people with distinct perspectives.

114

Be careful of 'Gold Plating' and keeping the project going beyond scope. Stick to the plan!

115

Success is to be measured not so much by the position that one has reached in life as by the obstacles which he has overcome.
~Booker T. Washington

116

Motivation means nothing compared to discipline. Discipline doesn't always feel good, it just works.

117

The ability to self-heal purposefully and naturally will go beyond any basic medications.

118

It is not sufficient to see and to know the beauty of a work. We must feel and be affected by it. ~Voltaire

119

Be open to learn from anyone that wants to give you their time and attention.

120

Minimize 'Default Fun' because it usually involves behaviors and activities that do not get us closer to our long term goals.

121

Sometimes it is better to be reminded then to be told.

122

It's pretty cool to see yourself on a video game.

~Rickie Fowler

123

Do not let the tech world diminish your faith.

124

If you can't imagine it you can't realize it. Think big.

125

Wisdom not shared is wisdom wasted.

126

You can search throughout the entire universe for someone who is more deserving of your love and affection than you are yourself, and that person is not to be found anywhere. You yourself, as much as anybody in the entire universe deserve your love and affection.

127

It is supposed to be tough. You are supposed to learn, move forward and gain the discipline required for the hard work required to reach any worthy goal.

128

We will most likely find the perfection of robots to make aspects of life more boring. Mistakes are inherently human and part of the journey.

129

Just because something doesn't do what you planned it to do doesn't mean it's useless.

~Thomas Edison

130

Simple is almost always better.

131

What would the super star, the super hero version of yourself do?

132

Growing slowly is still growing. Keep going.

133
The weak can never forgive. Forgiveness is the attribute of the strong.
~Mahatma Gandhi

134
Failure will never overtake me if my determination to succeed is strong enough.
~Og Mandino

135
Life is about personal growth and acquiring the ability to find some love in the things that seem unlovable.

136

When in true learning mode there is no such thing as failure. Keep going.

137

Be sure you always know how your developments will be used when you no longer control them.

138

Providing a judgment free environment for family, friends and colleagues to be heard may be the best gift you can give.

139

It's all about working to create the space where you have options.

140

Laugher is the medicine you take to help keep the need for real medicine away.

141

Give loyalty to those who give it to you.

142

Watch the way people treat you differently when you use "sir" or "Mam," especially those who might never be addressed that way by others.

143

No amount of riches will satisfy
the deepest questions in your soul.

144

The path to greater happiness is in
sacrifice.

145

If you don't act from the heart you might as well be a robot.

146

Build your communication skills with the same veracity you approach your technical work. If people can't understand the answer will always be "no."

147

Life is often more about context than content. Be sure to note the context you see and realize their maybe even be a larger one.

148

If you are trying to fulfill a dream then there is a good chance you are spending a lot of time by yourself. This doesn't mean you are lonely.

149

In an instant your 30 year career can be wiped out without even an error, so be careful what you actually do say and how it is said.

150
The real problem is not whether machines think but whether men do.
~B. F. Skinner

151
One machine can do the work of fifty ordinary men. No machine can do the work of one extraordinary man.
~Elbert Hubbard

152

Create a home that values integrity, communication, empathy, humility, gratitude, love and one that increases everyone's opportunity to laugh.

153

If you do not provide an alternative for whatever is broken people will create their own solution.

154

Think about how much code you can cut?

155

Hope is a waking dream.
~Aristotle

156

The best leaders and followers ask the best questions.

157

Many scientists think that philosophy has no place, so for me it's a sad time because the role of reflection, contemplation, meditation, self-inquiry, insight, intuition, imagination, creativity, free will, is in a way not given any importance, which is the domain of philosophers.
~Deepak Chopra

158

Seek happiness before you seek success or you may find neither.

159

Don't worry when you are not recognized, but strive to be worthy of recognition.
~Abraham Lincoln

160

Always be yourself, express yourself, have faith in yourself, do not go out and look for a successful personality and duplicate it.
~Bruce Lee

161
By 2030 500 million jobs will be wiped out globally. That's less than ten years, what are you going to do now to be ready?

162
Failure is a far better teacher than success. Early success might even be a motivation killer.

163
Be sure to reach out to your human colleagues and demonstrate some human qualities, every day.

164

Only do what you are great at, the others will just slow you down.

165

Move beyond the screens to create interaction. You have to find a subject that interests you enough to talk before you find the ones to connect with.

166

If you don't think your machine is that powerful, wait until something doesn't work right and stand back.

167

Shake your hand and smack you upside the head with your other...that is what technology does on a continual basis.

168

If you can't design it, build it. If you can't build it, fix it. If you can't fix it, get someone who can and search out another role.

169

Seek how to be happy before searching how to be successful. Not only is it the wrong order, it won't work in the end.

170

Treat the customer like you would treat a family member, that you actually like.

171

Don't get caught up being successful at things that don't matter.

172

If you don't move fast enough or refuse to take a risky leap forward, you will likely not get the prize.

173

What a great feeling to be there to witness someone experiencing great joy, especially when you had a part in it.

174

As human employees we must demonstrate our value or we will eventually be discarded by businesses like an old software version.

175
We can blame our mistakes on computers because they can't stand up for themselves.

176
Be sure that you ritualize success or you will never feel anything but a grind.

177
The human spirit must prevail over technology.
~Albert Einstein

178

Your internet security is, with all your good intentions, not enough. Find the next way to lock it down.

179

Look up at the stars and not down at your feet. Try to make sense of what you see, and wonder about what makes the universe exist. Be curious.
~Stephen Hawking

180

The irony is that humankind is spending so much time on trying to build the perfect human when it has already been done.

About the Author

Dave Gerber is President and founder of Synergy Development & Training, LLC. This innovative organizational solutions company specializes in helping businesses, government agencies, organizations, schools and individuals build leadership capacity, *use conflict* as an opportunity to increase performance and revenue, and reduce risk.

Dave helps individuals understand the impact of conflict and specializes in motivating others to increase and actualize their potential, regardless of industry as he has worked within many. He has worked with business owners, executives, managers, employees of all kinds, CEOs, military officers, the space and intelligence community, business development directors, doctors, lawyers, engineers, educators and more in a coaching, training, speaking and consulting capacity. He has hands-on experience working with over 10,000 training participants and students of all ages, races, backgrounds and ability levels.

Dave received a Bachelor's Degree in Sociology from Ithaca College in New York and a Master's Degree in Education from St. Joseph's University in Philadelphia. He also holds a Senior Executive Leadership certificate and a Leadership Coaching (ICF) certificate from Georgetown

University in Washington, D.C. He holds Executive Certificates in Negotiations, Leadership and Management from Notre Dame. He has multiple certificates in Workplace Conflict Processes, Workplace Mediation and Conflict Resolution, Commercial, Federal Workplace and Family Mediation from the Northern Virginia Mediation Service at George Mason. Dave has other professional certificates from Cornell, Notre Dame, Villanova, MIT and others. Dave brings over 20 years of varied conflict management, teaching, coaching and educator training experience as well. He has also written a dozen books with more out shortly.

Dave is also a certified Black Belt Krav Maga Instructor and Self-Defense coach. He is the founder and Chief Instructor of L.I.F.E. Line Self-Defense and Under Dog Krav Maga. He is also a certified Reiki Master and has created his own healing system. His desire to serve, support and lead leaders is limitless.

NOTES

NOTES

NOTES

NOTES

www.PreparingforRobots.com

NOTES

www.DaveGerber.info

NOTES

www.PreparingforRobots.com

NOTES

NOTES

www.PreparingforRobots.com

NOTES

www.DaveGerber.info

NOTES

www.PreparingforRobots.com

NOTES

NOTES

www.PreparingforRobots.com

NOTES

Made in the USA
Monee, IL
04 November 2026

67383028R00056